EXECUTIVE SUMMARY

TECHNOLOGY FORCES AT WORK

Profiles of Environmental Research and Development at DuPont, Intel, Monsanto, and Xerox

Susan Resetar

with

Beth E. Lachman

Robert Lempert

Monica M. Pinto

Supported by the

Office of Science and Technology Policy

RAND

SCIENCE AND TECHNOLOGY POLICY INSTITUTE

PREFACE

The environmental literature recognizes the importance of involving multiple stakeholders in the environmental policy development process. Stakeholders include a diverse set of individuals and organizations—local citizens and community groups, consumers, environmental groups, industry, individual companies, shareholders, all levels of government and tribes, etc. In turn, each of these will have various perspectives on environmental risk, priorities, costs and benefits, etc.

One area of environmental policy that has not received a lot of emphasis in the past is technology innovation. Because of this, there is limited information on how one of these key stakeholders, industry, views environmental research and technology innovation. This report summarizes information about the following:

- How research-intensive companies are rethinking investments in environmental technologies; where these companies are likely to invest, where they will not invest, and where opportunities for public-private sector partnerships are; and
- What federal policies the case-study companies would like to see to promote investments in environmental research and technology.

The work was sponsored by both the Environment and Technology Divisions of the White House Office of Science and Technology Policy.

The information contained in this summary is based on a series of interviews with senior environmental research and technology managers and environmental, health, and safety personnel in four research-intensive companies. The complete report that presents more detail on the results and the case studies of the companies is *Technology Forces at Work: Profiles of Environmental R&D at DuPont, Intel, Monsanto, and Xerox*, MR-1068-OSTP, 1999, by Susan Resetar with Beth Lachman, Robert Lempert, and Monica Pinto. The results of this study should be useful for federal, state, local, and tribal environmental and

R&D policymakers and scientists; industrial managers and planners; and university researchers.

Originally created by Congress in 1991 as the Critical Technologies Institute and renamed in 1998, the Science and Technology Policy Institute is a federally funded research and development center sponsored by the National Science Foundation and managed by RAND. The institute's mission is to help improve public policy by conducting objective, independent research and analysis on policy issues that involve science and technology. To this end, the institute

- supports the Office of Science and Technology Policy and other Executive Branch agencies, offices, and councils
- helps science and technology decisionmakers understand the likely consequences of their decisions and choose among alternative policies
- helps improve understanding in both the public and private sectors of the ways in which science and technology can better serve national objectives.

Science and Technology Policy Institute research focuses on problems of science and technology policy that involve multiple agencies. In carrying out its mission the institute consults broadly with representatives from private industry, institutions of higher education, and other nonprofit institutions.

This report is also available through RAND's web site. Inquiries regarding the Science and Technology Policy Institute or this document may be directed to:

Bruce Don
Director, Science and Technology Policy Institute
RAND
1333 H Street, N.W.
Washington, D.C. 20005
Phone: (202) 296-5000
Web: http://www.rand.org/centers/stpi/
Email: stpi@rand.org

CONTENTS

ACKNOWLEDGMENTS

The cooperation of the following individuals helped make this study possible. They gave of their time and ideas and provided thoughtful insights that enriched this report. Although the report attempts to reflect the conversations with participating companies accurately, any errors are the responsibility of the authors.

We greatly appreciate the time the following individuals spent describing their activities: John Carberry, DuPont's Director of Environmental Technology; Terry McManus, Intel's Assembly-Test Manufacturing Environmental Health and Safety Manager; Tim Mohin, Intel's Manager of Corporate Environmental Affairs; from Monsanto, Phil Brodsky, Vice President, Corporate Research and Environmental Technology, Earl Beaver, Director, Waste Minimization, and Laurence O'Neill, Director, Environmental Communications; and from Xerox, Allan E. Dugan, Senior Vice President, Corporate Strategic Services, Mark Myers, Senior Vice President, Corporate Research and Technology, Rafik Loutfy, Vice President, Strategy and Innovation for Corporate Research and Technology, Jack Azar, Director, Environment, Health, and Safety, Jim Cleveland, Manager, Quality, Technology and Engineering, Ralph Sholts, Manager, Recycle Technology Development, Stephen Dunn, Manager, Environmental Technology, Patricia Calkins, Manager, Environmental Products and Technology, Ronald Hess, Manager, Environmental Engineering Programs/Operations, and Joseph Stulb, Manager, Environmental Engineering Operations

David Rejeski, Council on Environmental Quality (formerly of OSTP), and Kelly Kirkpatrick of OSTP have both been extremely supportive and engaged clients throughout the entire research process. Lloyd Dixon and Ron Hess both gave thoughtful, timely technical reviews that have improved the report's quality. Katie Smythe's review of this executive summary ensured that the report's messages were clearly and comprehensively presented. In addition, we would like to thank Cynthia Cook for her informal review of the entire document and Nicole DeHoratius, Elisa Eiseman, and Deborah Sole for their reviews of the Intel, Monsanto, and DuPont cases, respectively. And Daniel Sheehan and Jerry Sollinger have contributed immensely to the report's presentation. The contents and conclusions are the sole responsibility of the authors.

EXECUTIVE SUMMARY

> In the United States, it takes 12.2 acres to supply the average person's basic needs; in the Netherlands, 8 acres; in India, 1 acre. . . . [I]f the entire world lived like North Americans, it would take *three planet Earths* to support the *present* world population. (Emphasis added.)
>
> —*Donella Meadows (1996).*

INTRODUCTION

The quotation above underlines the need to maintain economic growth without increasing—and preferably decreasing—the material and energy resources needed to achieve that growth. The world's population is increasing; its material and energy resources are not. Needed are new techniques that enable development without increased demand for those resources. In addition, new cost-competitive techniques to realize environmental benefits will help our industries remain competitive in the global marketplace.

Environmental technologies "advance sustainable development by reducing risk [of human health or environmental harm], enhancing cost-effectiveness [of achieving a level of environmental protection], improving process efficiency, and creating environmentally beneficial or benign products and processes." (NSTC, 1994, p. 9.)[1] While these technologies are not sufficient in themselves to achieve economic growth and improved quality of life without using more energy and material resources, they are necessary because the improvements will not occur without them. New technologies can potentially provide lower-cost means for achieving a given level of environmental protection. Thus everyone has a stake in fostering innovative environmental technologies. Unfortunately, past federal environmental policy did not emphasize technological innovation as a way to achieve better environmental performance at lower cost.

[1] Brackets are added clarifications. "Sustainable development" refers to the need to allow growth while balancing economic, environmental, and social needs now and in the future.

(OTA, 1994; EPA, 1992; EPA, 1993.) Furthermore, environmental policy did not necessarily accommodate differences in firms' behaviors toward environmental issues—one policy fit all (Rejeski, 1997). As a result of this simplification, the government does not know much about how a key environmental stakeholder—industry—views environmental research and technology innovation.[2]

This lack of information is regrettable, because industry not only provides the goods and services that have an enormous effect on the environment, but it also offers a tremendous resource for developing new environmental technologies. Industry is a font of talent, has access to even more, and funds about two-thirds of all research and development (R&D) performed in this country, outspending the federal government by $2 to every $1. However, as in the federal government, its R&D budgets are under pressure, albeit for different reasons. To retain a place in the fiercely competitive global market, firms must keep costs as low as possible, and R&D budgets do not escape cost scrutiny. Thus, it is important that R&D dollars get spent as effectively as possible.

The federal government has an opportunity to advance its own environmental goals by complementing the work of U.S. industry. However, to do so, federal policymakers must understand what environmental technologies industry invests in now, what it will invest in tomorrow, and why it makes these investments. A clear understanding in these areas will enable policymakers to guide federal spending, assist in deciding which public-private partnerships to form, and craft policies that stimulate the most effective environmental technology R&D in industry.

THIS STUDY: WHAT IT DOES AND HOW

This report attempts to fill in part of the information gap and increase policymakers' understanding of these issues by illuminating *emerging* environmental technology R&D trends in a limited number of industrial sectors. The study addresses two major research issues. First, it addresses how research-intensive companies are rethinking investments in environmental technologies; where these companies are likely to invest, where they will not invest, and where opportunities for public-private partnerships are. Second, it inquires into what federal policies the case-study companies would like to see to promote investments in environmental research and technology.

---

[2]The list of potential stakeholders includes a diverse set of individuals and organizations—local citizens and community groups, consumers, environmental groups, industry, individual companies, shareholders, and all levels of government and tribes. In turn, each of these will have various perspectives on environmental risk, priorities, costs and benefits, etc.

Ultimately, this information may help improve the federal R&D policies that promote environmental technologies. The many policy choices include direct federal investment, public-private research partnerships, federal support for high-risk technology demonstrations, use of federal laboratories' capabilities and resources, research and experimentation tax credits, and federal-state collaboration on research efforts. Other policies that influence environmental technology R&D investments include environmental regulations, product liability laws, green labeling programs, federal and state procurement criteria, foreign aid and technology assistance, education and training investments, and programs to collect and disseminate environmental information.[3]

To carry out this research, we performed four case studies of firms in different industries. The four companies identified as leaders in quality R&D processes and the treatment of environmental issues were DuPont, Monsanto, Intel, and Xerox.[4] These are large, multinational manufacturing organizations with significant R&D investments that represent the chemicals, biotechnology, and electronics sectors.[5]

Each company interviewed has some level of experience with a range of federal programs—notably the Environmental Protection Agency's (EPA's) Remediation Technologies Development Forum, the Department of Energy's (DoE's) Industries of the Future Program, the Department of Commerce's (DoC's) Advanced Technology Partnerships, cooperative R&D agreements with federal laboratories, and the EPA's regulatory reinvention Project eXcellence and Leadership (Project XL).

OBSERVATIONS

We have identified 10 primary messages in two broad categories. The first set of observations covers environmental technology investments in the case-study companies. The second category of observations discusses the kinds of federal policies that can potentially be employed to stimulate these investments. Our

---

[3]This is not an exhaustive list of potential federal policy tools, but it does provide an illustrative overview. For more specifics on federal policies, see NSTC (1994) or OTA (1994).

[4]Leaders were identified by reviewing mentions in the R&D and environmental literatures on innovativeness, reputation, and quality management processes; participation in voluntary programs and environmental management initiatives; and environmental and management award-winners combined with the subjective judgment of select industry experts. At this time there is no method to rigorously and irrefutably quantify clear consensus on who is "the leader"—but these companies are among those with a novel perspective.

[5]These sectors were chosen somewhat arbitrarily based on interest from an ad hoc working group of government personnel and a qualitative assessment that the technological maturity and environmental issues among these sectors were different.

observations are based on the interviews and the literature combined with our judgment.

Regarding Environmental Technology Investments

Investments in Environmental Technology R&D Are Substantial But Unquantified. The case studies suggest that investments in all categories of environmental technologies are "large," but only one quantitative estimate of "large" was provided. A couple of published estimates range from 1 percent to 13 percent of all R&D is devoted to pollution-control devices (one category of all environmental technology). While another estimates that 50 percent of R&D has an environmental, health, and safety component. These estimates were calculated in different time periods, employ different definitions of environmental technology, and were based on different samples.

Systematically Collected Quantitative Data on Industrial Investments in Environmental Technology R&D Will Improve Policymaking. Without systematically collected quantitative information on the amount of industrial R&D investment that has an environmental component, where these investments are being made, and how these investments are changing (in response to markets and policies), only a limited understanding of the effectiveness of future public policies can be gleaned.

Leading Companies Invest in Environmental Technologies to Improve the Resource Efficiency of Their Products and Manufacturing Processes Because It Is Cost-Effective to Do So. Our sense is that many improvement opportunities still exist. These opportunities may involve either organizational or technology innovations. The companies are also actively tracking global trends in resource scarcity, environmental regulations, voluntary product standards, and customer environmental priorities and needs so they can respond rapidly to emerging markets. Because much of this research deals with proprietary knowledge about products and processes, extensive collaboration is less likely than it is for other technology areas.

Companies Rely on a Rich Science and Technology Base for Environmental Technology Innovations. All of the companies relied on universities for new knowledge and to provide a trained workforce. They also used smaller technology-based companies for niche capabilities to complement in-house research. Both of these institutions are especially important for radical change because they provide knowledge and capabilities outside the areas traditionally emphasized within a firm.

All the Companies Had Experience with an Innovation That Led to Additional Innovations. For some, these were new applications of the technology. In others, they were refinements to existing features. In each case, the experience led to additional environmental improvements. For completely new technologies, experience helped the companies address changing market and customer needs, which often develop at the same time as the technology.

On Policies to Enhance Environmental Technology Investments

The Companies Look to the Government for Better Information and Leadership About Environmental Priorities. Innovation and diffusion, especially radical innovation, can take a long time—decades or more. However, determining the right time to invest involves predictions about markets, technologies, sociopolitical conditions, and regulations. Being wrong can be costly. But clear signals and leadership from the government can reduce some of this uncertainty. Right now the strongest signals to these firms regarding global, national, and local environmental priorities are regulations and customer preferences. More information and data—on the full cost of materials and energy use, energy and material flows, chemical toxicity, etc.—will ensure that informed decisions are made. Better information on what is an environmentally preferable or sustainable product gives industry and its customers the opportunity to make better choices and will stimulate more investment.

Environmental Regulations Have Clearly Influenced the Firms. The Toxic Release Inventory may have stimulated these companies to look at emissions as opportunities to save money; the time and expense of gaining environmental permits is causing some to practice pollution prevention; and emissions controls, hazardous waste management, and other regulations are stimulating them to rethink their own and their customers' material and energy flows. Environmental regulations can create markets for environmental technologies by changing the cost structure of emissions.[6] However, the extent to which regulations and not other management practices, such as total quality management, ISO 14000, or supplier management, influence environmental R&D investments is unclear from the information collected in these interviews.

---

[6]Environmental regulations have had negative and positive effects on innovation. The literature discusses how environmental regulations can add another element of uncertainty to investment decisions and limit long-term innovation because of uncertainty regarding the form of the regulation, its enforcement, the administrative burdens of verifying performance and modifying permits, market segmentation resulting from differing standards, and the liabilities associated with potential performance failure. Some of these aspects of the regulatory process reduce the benefits to technology providers of investing in new technologies, and some are disincentives for regulated companies to be the first adopter of new compliance technology. The aversion to being first adopter of new compliance technology was expressed in our interviews as well. In contrast, as companies link environmental issues to corporate strategy, competitive advantage will encourage some companies to be an early adopter of new technology.

Well-Managed and Scientifically Rigorous Environmental Regulatory Practices Are an Important Policy Tool. They can be used to negotiate the risk and uncertainty associated with new environmental technologies and thereby speed their diffusion. From the companies' point of view, public confidence in the regulatory process is as important as the scientific practices employed because it improves public acceptance of new technologies and new approaches.

Strong Intellectual Property Rights May Not Necessarily Be Appropriate for Environmental Technologies. The case-study companies would like to see the federal government work internationally to ensure global enforcement of patent laws to protect intellectual property. However, other considerations need to be balanced with that enforcement. Because environmental technologies have a large public-good aspect, rapid diffusion is desired to more quickly realize the environmental benefits of the new technology. Diffusion may also spur additional innovations, leading to new environmental and cost benefits. While strong intellectual property rights create incentives for companies to invest in R&D to generate new technologies, at some point strong property rights could slow diffusion by limiting access to a new environmental technology or by raising its price. (Widespread use of licensing can mitigate this problem.) As a result, new systems for protecting intellectual property must balance these somewhat competing issues.

Effective Federal Policies to Promote Environmental Technologies Will Require Multiple Policy Tools. Government efforts cannot simply rely on a single tool, such as environmental regulations. While regulations are important, the landscape of environmental technology R&D is complex, and no single tool will sufficiently foster the full range of environmental technology R&D investments. Federal investments in science improve the knowledge base for environmental priority-setting and stakeholder processes will help create consensus. Support for university-based research may foster dramatically new technological options as well as train the next generation of researchers and engineers that industry will rely on. Raising consumer awareness will increase demand for products that have improved environmental performance and help with environmental priority-setting. Public-private partnerships may help leverage funds to address common technology issues or may be effective means to build consensus among stakeholders. These policies address different elements of the innovation process, all of which are important to new technology development and deployment.

The remainder of this summary elaborates on the information gained through our interviews with the four companies. It provides more detail on the lessons drawn from our discussion, and it presents the companies' views of what they would like to see in terms of government environmental technology policy.

CROSS-CUTTING LESSONS LEARNED FROM THE CASE STUDIES

The first question that this study addresses is how research-intensive companies are rethinking investments in environmental technologies. Specifically, we examined where these companies are likely to invest, where they will not invest, and where opportunities for public-private partnerships are.

Where the Companies Are Likely to Invest

The four case-study companies are most likely to invest in technologies that increase product and process resource efficiency, create more environmentally benign products, improve manufacturing yield or reduce emissions, and meet customer product requirements. They recognize that environmental issues touch a majority of their R&D investments, although they are not generally the primary reason for the investment. The innovation process they describe emphasizes the "demand pull" of technology to meet customer needs, community concerns, market trends, regulations, or their corporate environmental goals. All the companies interviewed are actively monitoring trends in environmental regulatory policy, customer preferences, customer needs, and resource constraints to determine technology investments.

Because environmental goals are only one of many corporate objectives that influence the R&D portfolio, leading companies integrate these issues into corporate strategic planning to improve their visibility and to increase the understanding of decisionmakers throughout the corporation on how environmental issues influence corporate profits. They are actively seeking opportunities through which addressing environmental concerns makes good business sense. (Not surprisingly, the technologies that more readily attract R&D investment also relate to other corporate objectives of profitability, cost reduction, and market access.) In many cases, greater integration and cooperation on environmental issues among manufacturers and their suppliers, distributors, and customers is occurring. Because of the anticipatory nature of strategy, a key question for companies becomes timing—or when to invest. (See text boxes for insights into how the four case-study companies approach environmental technology.)

Where the Companies Are Unlikely to Invest

Industries expressed less interest in remediation, monitoring, and control technologies. While often necessary to meet regulatory requirements, control and remediation technologies are less likely to meet other corporate objectives and as such are generally not cost-effective investments. As a result, the case-study

DuPont: Seeking to Solve Its Customers' Environmental Problems

DuPont's investment priorities are yield improvement, sustainable or environmentally preferable products and services, reuse and recycle, and control and abatement technologies. DuPont executives have stated that biotechnology, as applied to the chemical and life sciences businesses, has the potential "to replace raw material, energy, and capital-intensive processes with low-temperature, low-pressure, zero-waste, and cheaper routes to products important to society." (Holliday in Reisch, 1998.) Approximately 95 percent of all DuPont R&D investments have at least a modest environmental aspect; 65 percent have a large environmental aspect; and 15 percent are exclusively environmental.[a] DuPont integrates environmental issues into its strategic planning process by ensuring customer environmental needs are addressed, applying a World Business Council for Sustainable Development sustainability planning template, and following its own Safety, Health, and Environment Vision statement during discussions among senior managers and planners. In 1996, DuPont executives estimated that the next 50 percent improvement in waste reduction per pound of product will save $3 billion to $5 billion.[b]

Environmental technology R&D planning at DuPont has moved away from a cost containment perspective to one that focuses on market-driven opportunities and sustainability. One major thrust at DuPont is to anticipate and solve customers' environmental needs. For example, DuPont is applying its own hazardous-material-handling experience and capabilities to its customers' operations, either giving them advice on dealing with hazardous materials they supply, providing the handling services, or installing the material in an industrial component, relieving the customer of handling the hazardous material at all. Another emphasis is product research and technology that may make DuPont the supplier of choice because of a reputation for environmental excellence or because its products meet environmental standards required by customers. For example, DuPont is developing an environmentally benign process (it requires no heavy metals, petroleum, or toxic chemicals, and the effluent is biodegradable) to manufacture polyester intermediates using glucose from corn starch (a renewable) and enzymes from a microorganism created through recombinant DNA. The microbial byproduct can by used as animal feed, and DuPont has a patented process to unzip these molecules for recycle (Krol, 1997; Holliday, 1998).

---

[a]Exclusive investments include control and remediation technologies; design, project engineering, and installation for abatement equipment; development of Freon® alternatives, and product and process changes for the Montreal Protocol. Largely related investments are those that improve quality or resource efficiency. For example, development and engineering for process improvements that improve first-pass yield. Modestly related are those activities that might improve product quality with some minor environmental improvement.

[b]Personal communication with John Carberry indicated that this number includes material cost savings, disposal cost savings, and lost revenue opportunities. The loss in potential revenues for input materials is the largest portion of this estimate by far. Waste per pound of product will be reduced with the range of DuPont's environmental technology investments—yield improvements, sustainable products, reuse/recycle, zero emissions, and co-product development technologies (DuPont, 1996, quoting Paul Tebo, Vice President, Safety, Health, and Environment).

companies sought to limit the amount invested in these technologies and invested only when necessary to comply with regulations (and if outside sources were not available). The companies hoped that the pollution-prevention orientation of the technologies of interest (such as yield improvement technologies and emissions reduction technologies) would to some extent lessen future requirements for remediation and control technologies.

The Role of Public-Private Partnerships

All four case-study companies were involved with public-private research partnerships, and most wanted them to continue. However, the companies' views were mixed regarding priorities for the focus of research partnerships. Some felt that such partnerships were appropriate for remediation and end-of-pipe pollution control technologies, because these are more readily generic or precompetitive technologies.[7] For others, remediation and control technologies were no longer a priority. These firms were interested in collaborating on recycling and remanufacturing, yield improvement, energy efficiency, or emissions reduction technologies. Two companies specifically mentioned that the activities of public-private partnerships were used for noncritical technology enhancements or alternative technological approaches. This suggests that targeted public-private partnerships can be useful to industry if used judiciously.

CASE-STUDY COMPANIES' RECOMMENDATIONS FOR GOVERNMENT

The case-study firms see clear roles for the federal government in environmental technology R&D, and they offered a number of recommendations. Before discussing these in detail, we mention two caveats. First, the industrial sector is much more diverse than the sample represented by the companies involved in this study, which are not necessarily generalizable to the entire industrial sector. Second, these recommendations for federal policy reflect the perspective of a subset of only one stakeholder—industry—and, as such, may not be appropriate for federal action when all stakeholder interests are taken into account. Nevertheless, the case-study companies' recommendations provide insight into the preferences of an important stakeholder in environmental technology policy.

---

[7]Antitrust legislation prohibits firms from engaging in collaborative research on competitive technologies.

Recommendations Common to Several Companies

Several themes—provide leadership, invest in science and technology, develop markets, and protect intellectual property—came up repeatedly in our discussions.

Provide Leadership. All the companies interviewed wanted the federal government to provide leadership on national environmental technology priorities. Improved information based on scientific knowledge will aid the companies' decisionmaking. As companies go beyond compliance, because of cost savings or developing markets, they will face decisions about product features and content, technological options, and emissions trades, among others.[8] Federal leadership on national environmental priorities, operationalizing sustainability and systems thinking, consensus-building using science, and data collection

Intel: Prevention to Enable Rapid Change

Intel is known for introducing rapid series of incremental product improvements—its motto is "quick or dead." As a consequence, product and process design occurs simultaneously. Every two years, production facilities are completely retooled. Environmental regulations are constraining and costly for this pace of innovation, so Intel is investing in pollution-prevention technologies with the goal of reducing emissions below levels that require an environmental permit. This approach will save the company money by avoiding the administrative expenses of obtaining permits. Plus, it avoids losing opportunities to generate revenue because of delays to manufacturing process changes. Intel also invests in process technologies to improve water conservation for both ultrapure water and wastewater because of community concerns. Its Chandler, Arizona, fabrication facility recycles wastewater externally, enabled by city infrastructure. Other technology areas of interest to Intel include chemical-use reduction and solid-waste conservation.

Intel participates in the EPA's Project XL experiment so that it may gain process flexibility. Public comments on the Intel final project agreement for the Project XL experiment suggest areas for environmental R&D. The inability to define superior environmental performance precisely, the lack of toxicological data and risk assessment methods, and the inability to closely monitor actual environmental performance—particularly hazardous air pollutant emissions—have contributed to differing interpretations of Intel's plan to achieve superior environmental performance and its enforceability.

---

[8]The authors wish to note that not all firms will choose to move beyond the requirements established by environmental regulations. Just how widespread this behavior may be is beyond the scope of this study. Other authors discuss the various strategies firms may take toward environmental issues ranging from noncompliance to leadership. For example, see Roome (1994) and Chatterji (1993).

can improve industrial environmental decisionmaking and encourage the requisite R&D investments.

Invest in Science and Technology. The companies overwhelmingly felt that continued support for a strong science and technology base was an important role for the federal government. All the companies have ties to academia, and it is clear that access to university-based research helps the companies with their research agendas by either augmenting internal research or gaining unique expertise that did not exist within the company. They also looked to the universities to provide a scientifically trained workforce. Public education was also considered important to facilitating acceptance of new technology and environmentally preferable products as well as to helping equip local communities to establish environmental goals and priorities. A few were interested in support for the national laboratories for specialized skills or facilities. Scientific and technology areas mentioned include biotechnology, chemistry based on biological analogies, information technology, nanotechnology, energy-efficiency technologies, and development of sustainable products.

Develop Markets. Because the companies' first priority is to meet customer requirements, policies that create "market pull" are a straightforward way to draw

Xerox: Toward an Industrial Ecology, Closing Material and Energy Flows

Product-related technologies in the areas of energy efficiency, chemical and physical emissions, natural resource conservation, and waste management receive approximately equal amounts of investment at Xerox. Customer requirements and Xerox's own asset recycle management initiative, which remanufactures and reuses copiers and other equipment, directly link these environmental investments to the bottom line. Xerox has a global tracking system for consumer and regulatory trends. Information from this system is presented to the research centers, product teams, and manufacturing units to integrate environmental issues into R&D planning. Xerox estimates show the company is avoiding *hundreds of millions of dollars* in cost because of its environmental initiatives. Because equipment is reused, savings in virgin material and energy are also achieved.[a] Early manufacturing facility planning and pollution prevention through material selection and process design are two tactics for reducing the complexity and time required by the environmental permitting process. However, with rapid product innovation, environmental planning for permitting is quickly becoming a constraint on process cycle times for several product lines (in some cases it is already the primary constraint).

---

[a]Some of the savings in material and energy used to produce virgin material and parts will be offset by the material and energy required to transport, sort, clean, and refurbish returned products. A detailed life-cycle analysis of the environmental impacts of remanufacturing is not available.

environmental investment. A couple of ideas suggested creating markets for environmental products to attract additional investment were affirmative procurement and labeling environmentally benign products and processes.

Protect Intellectual Property Rights. As industry substitutes information for tangible resources, the protection of intellectual property carries even greater importance to companies investing in R&D. Industry looks to government to be aggressive in defending U.S. intellectual property in the global marketplace.

Monsanto: Substituting Information for Material Resources

Monsanto seeks to create market value without necessarily using more material and energy resources. One example of how Monsanto might add value without adding resources would be to create products that can be used for multiple purposes—for example, a plant that bears fruit and whose leaves could be used to create biodegradable plastic. Monsanto executives believe that creating products to improve quality of life around the world—such as those that improve food supplies or lead to better health—will, in turn, generate new markets for its products. Monsanto also seeks to address markets created from anticipated shortages in natural resources, such as water and energy. Much of Monsanto's R&D has environmental implications, although these investments might not be classified as such. For example, a cost analysis for a rubber stabilizer showed that 1.5 pounds of waste was generated per pound of product—money lost to Monsanto. After some R&D investment, the waste generated was reduced to 0.1 pound per pound of product. While this effort was categorized as a cost-reduction effort, environmental benefits were clearly achieved as well. Environmental issues will be integrated into strategic planning and decisionmaking with a set of analysis tools that address product sustainability, full cost accounting, and life-cycle material and energy flows (based on Monsanto's understanding of sustainability).

Monsanto has reinvented itself as a biotechnology firm. Such radical change takes a long time. Monsanto began investing in biology-based product research in the early 1970s. At that time, chemistry was king at Monsanto, and research focused on chemical solutions to crop management and process improvements. To transform itself, the company turned to university scientists and acquired or allied with small biotechnology companies, seed companies, pharmaceutical companies, and others to build the expertise and capability required to develop a host of biotechnology-based products in agriculture, pharmaceuticals, and animal health. Nearly 25 years after its initial investment in biology, Monsanto has introduced several new products based on its research investments and plans to apply its expertise in biotechnology to sustainability.

Other Recommendations

The following recommendations either were mentioned by only one company or received less emphasis during the interviews. These include improving regulatory policy to allow regional approaches and priority-setting, incorporating performance-based criteria across all media (air, land, and water), and encouraging pollution prevention. Other suggestions included increasing funding to regulatory agencies (EPA, FDA, USDA) to attract and maintain high-quality staff, to ensure public confidence in the regulatory process, and to bring new scientific discoveries into the regulatory process[9]; developing environmental and technology policies that help companies operate in a global economy; using federal investments to develop improved monitoring technologies and measurement standards; and funding science and technology for sustainable products, such as economically viable ways to collect, sort, clean, and disassemble materials at the molecular level to be able to make recycled material.

These recommendations represent the preferences of four innovative companies. Federal policymakers must balance these preferences with the needs of other industrial members, other stakeholders, and the cost-effectiveness of various policy tools. While markets, customer preferences, and profits are the preeminent drivers of investment, these suggestions by the case-study companies illustrate how government has had, and can have, an important role in fostering environmental technology investments. If we are to sustain economic and population growth without further jeopardizing human and environmental health, technology innovation may help achieve the orders of magnitude improvement needed. Sustainability is clearly an idea that industry leaders are wrestling with. Jack Krol (1997), former DuPont CEO outlined three challenges for the chemical industry. These challenges are value creation, technology, and sustainability. These challenges transcend the chemical industry. They are challenges for *all* industry, *all* government, and *every* citizen.

---

[9]Note that there may be ways to achieve these goals other than by increasing funding to these agencies, such as reallocating the distribution.

REFERENCES

Chatterji, Deb, "R&D Management and the Environmental Imperative," mimeo; British Oxygen, 1993.

DuPont, "Profitable Growth Will Drive Success," news release, January 26, 1996.

Environmental Protection Agency, *Report and Recommendations of the Technology Innovation and Economics Committee: Permitting and Compliance Policy: Barriers to US Environmental Technology Innovation,* EPA 101-N-91-001, National Advisory Council for Environmental Policy and Technology, Washington, D.C., January 1991.

_____, *Improving Technology Diffusion for Environmental Protection: Report and Recommendations of the Technology Innovation and Economics Committee,* EPA 130-R-92-001, Washington, D.C., October 1992.

_____, *Transforming Environmental Permitting and Compliance Policies to Promote Pollution Prevention: Removing Barriers and Providing Incentives to Foster Technology, Innovation, Economic Productivity, and Environmental Protection,* EPA 100-R-93-004, National Advisory Council for Environmental Policy and Technology, Washington, D.C., April 1993.

_____, *Intel Final Project Agreement Response to Comments,* 1996. Available at http://www.epa.gov.

Hart, Stuart, "Beyond Greening: Strategies for a Sustainable World," *Harvard Business Review,* January-February 1997, pp. 66–77.

Holliday, Chad, President and CEO of DuPont, remarks at the Biotechnology Panel Discussion of the World Economic Forum, Davos, Switzerland, February 3, 1998.

Krol, Jack, "The Chemical Industry—Indispensable in the 21st Century," speech presented to Pittsburgh Chemical Day, Pittsburgh, Pa., April 15, 1997.

Meadows, Donella, "Our Footprints Are Treading Too Much Earth," *Charleston Gazette,* April 1, 1996, in Hart (1997), p. 68.

National Science and Technology Council, *Technology for a Sustainable Future,* Washington, D.C.: Government Printing Office, 1994.

Office of Technology Assessment, *Industry, Technology, and the Environment: Competitive Challenges and Business Opportunities,* OTA-ITE-586, Washington, D.C.: Government Printing Office, January 1994.

Reisch, Marc S., "Holliday Blazes His Own Trail," *Chemical and Engineering News,* December 14, 1998, pp. 21–25.

Rejeski, David, "Forum: Getting Into the Swing," *Technology Review,* January 1997, pp. 56–67.

Roome, Nigel, "Business Strategy, R&D Management and Environmental Imperatives," *R&D Management,* Vol. 24, No. 1, 1994.